THE
CONCISE ATLAS
OF THE
STARS

A FIREFLY BOOK

Published by Firefly Books Ltd. 2005
Originally published as: *Le petit atlas des étoiles*

Copyright © 2005 / Larousse, Paris.

First printing

Publisher Cataloging-in-Publication Data (U.S.)
Brunier, Serge.
     Concise atlas of the stars/Serge Brunier ; constellation photography by Akira Fujii.
[64] p. : col. ill., maps ; cm.
Includes index.
Summary: A guide to 15 of the most important constellations visible in the Northern Hemisphere. Also includes sections on understanding the universe, the brightest stars and color photos throughout.
ISBN 1-55407-075-9
1. Stars — Atlases. 2. Northern sky (Astronomy) — Charts, diagrams, etc. 2. Constellations — Charts, diagrams, etc. I. Fujii, Akira, 1941– ill. II. Title.
523.80223 22   QB65.B8713 2005

Library and Archives Canada Cataloguing in Publication
Brunier, Serge
     The concise atlas of the stars / Serge Brunier ; constellation photography by Akira Fujii.
Translation of: Le petit atlas des étoiles.
Includes index.
ISBN 1-55407-075-9
1. Stars — Atlases. 2. Constellations — Charts, diagrams, etc. 3. Northern sky (Astronomy) — Charts, diagrams, etc. I. Fujii, Akira, 1941– II. Title.
QB65.B7813 2005   523.8'022'3   C2005-900493-2

Published in the United States by
Firefly Books (U.S.) Inc.
P.O. Box 1338, Ellicott Station
Buffalo, New York 14205

Published in Canada by
Firefly Books Ltd.
66 Leek Crescent
Richmond Hill, Ontario L4B 1H1

Printed in Malaysia

This book is dedicated to my dear mother, who, on clear nights, searches the sky for the telltale twinkling of Arcturus in Bootes, Deneb in Cygnus and Betelgeuse in Orion.

Credits and Acknowledgments

I would like to take this opportunity to extend a warm thank you to Marie-Pierre Levallois and Stéphane Aubin, without whom *The Concise Atlas of the Stars* would never have seen the light of day.

I would also like to thank those who accompanied me along the path of the Milky Way and who helped me find my way in this infinite powdering of stars: Catherine Delprat, Valérie Herman, Nathalie Bucsek, Roselyne Messager, Marie Dekerle, Isabelle Goulhot, Emmanuel Chaspoul, Laurence Lebot, Laurent Blondel and Marie-Astrid Bailly-Maître.

Management
Marie-Pierre Levallois

Editorial Direction
Catherine Delprat

Editing
Nathalie Cornellana, assisted by Manuela Laué

Artistic Direction
Emmanuel Chaspoul, assisted by Sophie Compagne

Artistic Design
Marie-Astrid Bailly-Maître

Layout
Michel Delporte

Maps
Laurent Blondel, Patrick Roth

Production
Nicolas Perrier

Translation
All Languages Ltd.

THE
CONCISE ATLAS
OF THE
STARS

Serge Brunier

Constellation photography by Akira Fujii

FIREFLY BOOKS

READING THE NIGHT SKY

T beautiful clear night has once again enticed you outdoors to stargaze. But tonight is different. Far from the city lights, the sky has a slightly unsettling presence. Here, toward the west, a bright star draws your attention. And there, in the south, a few twinkling stars outline a huge geometric shape against the dark sky. But how can you find your way around in this multitude of stars? How to recognize the constellations? In your exploration of the fascinating beauty of the sky you feel overwhelmed and you can't get your bearings. It seems impossible to name the heavenly bodies and, especially, to describe them. Twenty years of sitting "face to face" with the sky have led to the creation of *The Concise Atlas of the Stars* to help answer these questions and many more. Since classical celestial cartography makes reading the sky a complex task, we have removed everything that is not directly related to stargazing — each star map is reduced to the essential. Reproduced on transparencies, simply lay the diagrams over the beautiful photographs taken by Japanese astronomer and astrophotographer Akira Fujii to find your way around the starry sky.

It would take more than a lifetime to explore the entire sky, and such an endeavor is far beyond the scope of this book. As such, we have concentrated on only 15 of the total of 88 constellations. However, they are those easiest to observe on a clear summer night, and among them you will find the brightest and most interesting celestial objects. For each we have developed an "ID card" that outlines their specific features as we know them today.

The Concise Atlas of the Stars begins with the most famous constellation, the Big Dipper. Once you have found it, it will help you explore the dome of the sky throughout the summer months by following the slow, apparently clockwise movement of the stars from the early evening, through the middle of the night and into dawn. Although most of the constellations described in this book are visible in the northern hemisphere, some of them — Sagittarius and Scorpius, for example — are much easier to see in the southern United States and in the tropics. Finally, a few magnificent constellations, such as the Southern Cross, the Centaur and Carina, are only visible from tropical or southern latitudes.

The Milky Way, our galaxy, alone contains more than a thousand billion stars, and the visible universe contains more than a hundred billion galaxies. *The Concise Atlas of the Stars* will not provide a complete and definitive mapping of every star in the universe, but it will provide simple signposts for identifying the most beautiful celestial summer landscapes. If these pages help you recognize some stars in the sky the next time the night is clear, and if, night after night, you become more familiar and at ease with the constellations, then we have achieved our goal.

SERGE BRUNIER

HOW TO USE YOUR ATLAS

UNDERSTANDING THE UNIVERSE

OBSERVING 15 EXCEPTIONAL CONSTELLATIONS

⬤ Constellation
visible everywhere

⬤ Constellation visible from the
northern hemisphere

⬤ Constellation visible from the
southern hemisphere

Name of the constellation

Best time to view it

Its history and features

Main stars or interesting
objects (clusters, nebulas,
galaxies, etc.) that form
the constellation

The star's luminosity
compared to the
Sun's (Sun = 1)

Minimum requirements
for viewing (naked eye,
binoculars, telescope,
observatory)

The star's distance is
given in light-years (ly);
one light-year equals
10,000 billion kilometers

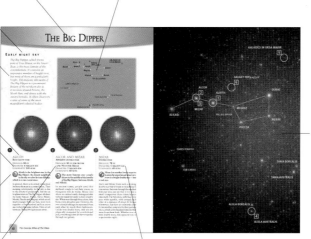

Map of the constellation
and surrounding area

Photo of the constellation:
to identify the main stars
or interesting objects that
are part of a constellation,
place the transparency
of the corresponding
diagram over the photo.

The dimensions or diameter of the star. In the case of a double star,
the dimension and brightness given are those of the main star.

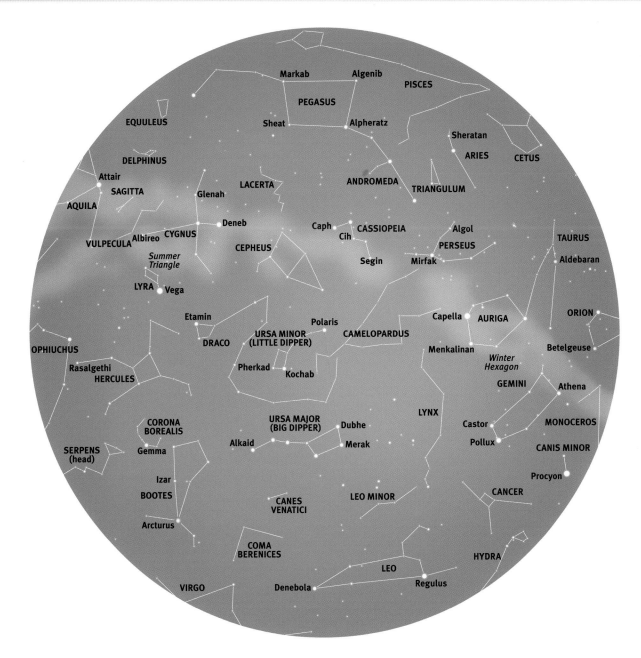

THE 88 CONSTELLATIONS OF THE SKY

In bold, the 15 constellations of *The Concise Atlas of the Stars*

Andromeda	Auriga	Canis Minor	Circinus	Corvus	Equuleus
Antlia	*The Charioteer*	*The Lesser Dog*	*The Drawing*	*The Raven*	*The Little Horse*
The Pump	Bootes	Capricornus	*Compasses*	Crater	Eridanus
Apus	*The Herdsman*	*The Sea Goat*	Columba	*The Cup*	*The River*
The Bird	Caelum	**Carina**	*The Dove*	**Crux**	Fornax
of Paradise	*The Chisel*	*The Keel [of Argo]*	Coma Berenices	*The Southern Cross*	*The Furnace*
Aquarius	Camelopardus	**Cassiopeia**	*Berenice's Hair*	**Cygnus**	Gemini
The Water Carrier	*The Giraffe*	**Centaurus**	Corona Australis	*The Swan*	*The Twins*
Aquila	Cancer	*The Centaur*	*The Southern*	Delphinus	Grus
The Eagle	*The Crab*	Cepheus	*Crown*	*The Dolphin*	*The Crane*
Ara	Canes Venatici	Cetus	Corona Borealis	Dorado	**Hercules**
The Altar	*The Hunting Dogs*	*The Whale*	*The Northern*	*The Swordfish*	Horologium
Aries	Canis Major	Chamaeleon	*Crown*	Draco	*The Clock*
The Ram	*The Greater Dog*	*The Chameleon*		*The Dragon*	

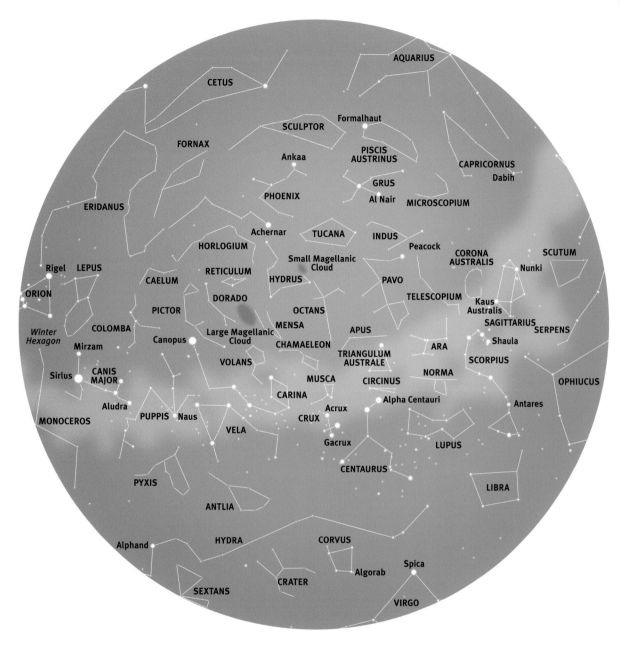

Hydra *The Water Snake*	Lupus *The Wolf*	Octans *The Octant*	Piscis Austrinus *The Southern Fish*	Scutum *The Shield*	**Tucana** *The Toucan*
Hydrus *The Water Snake (male)*	Lynx *The Lynx*	Ophiucus *The Serpent Bearer*	Puppis *The Stern [of Argo]*	Serpens *The Serpent*	**Ursa Major** *The Great Bear (The Big Dipper)*
Indus *The Indian*	**Lyra** *The Lyre*	Orion *The Hunter*	Pyxis *The Compass*	Sextans *The Sextant*	Ursa Minor *The Lesser Bear (The Little Dipper)*
Lacerta *The Lizard*	Mensa *The Table*	Pavo *The Peacock*	Reticulum *The Reticule*	Taurus *The Bull*	Vela *The Sails [of Argo]*
Leo *The Lion*	Microscopium *The Microscope*	**Pegasus**	Sagitta *The Arrow*	Telescopium *The Telescope*	Virgo *The Virgin*
Leo Minor *The Lesser Lion*	Monoceros *The Unicorn*	**Perseus**	**Sagittarius** *The Archer*	Triangulum *The Triangle*	Volans *The Flying Fish*
Lepus *The Hare*	Musca *The Fly*	Phoenix *The Phoenix*	**Scorpius** *The Scorpion*	Triangulum Australe *The Southern Triangle*	Vulpecula *The Fox*
Libra *The Scales*	Norma *The Square*	Pictor *The Easel*	Sculptor *The Sculptor*		
		Pisces *The Fish*			

THE DISTANT UNIVERSE

*H*undreds of galaxies up to 10 billion light-years away appear in this image of the distant universe, taken by the Hubble Space Telescope. The oldest galaxies appeared after the Big Bang an estimated 15 billion years ago. Since then, cosmic expansion has been causing them to drift further and further apart.

AN ISLAND IN SPACE

*F*loating in empty space, separated from each other by several million light-years, the galaxies are the cells that make up the universe. The youngest galaxies, such as galaxy M83 in Hydra, pictured here, are still generating stars in the vast nebulas that line their pinwheeling spiral disks.

OUR GALAXY

*T*he Sun, our star, is immersed in a giant spiral
galaxy, the Milky Way. In this image, only the
central core of the galactic disk appears, veiled by
clouds of gas and dust. This photograph of the galactic
center makes an interesting comparison with the image
of galaxy M83 on the previous pages.

THE MILKY WAY

*T*his photograph is an extreme close-up of a tiny section of the Milky Way pictured on the previous pages. It shows several million of the stars located in the heart of the galaxy some 10,000 light-years from the Sun. Astronomers estimate that the Milky Way contains more than a thousand billion stars.

THE BIG DIPPER

THE EARLY NIGHT SKY

The Big Dipper, which forms part of Ursa Major, or the Great Bear, is the most famous of the constellations. It contains an impressive number of bright stars, but none of them are particularly bright. The majestic silhouette of the Big Dipper is a prominent feature of the northern sky as it revolves around Polaris, the North Star, and shines with the aurora borealis. It offers observers a view of some of the most magnificent celestial bodies.

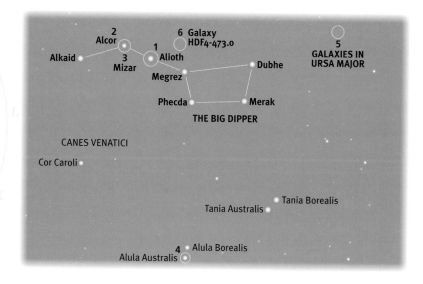

1

ALIOTH
BLUE GIANT STAR

DISTANCE: **82** LY
DIAMETER: **6 MILLION KM**
LUMINOSITY: **100 SUNS**

Alioth is the brightest star in the Big Dipper. Its closest neighbors in the sky are also its true siblings, which is a rare occurrence.

In general, there is no actual connection between the stars in a constellation. Their seeming relationship in the sky is due to the effects of perspective. But the six brightest stars in the Big Dipper all share the same history. Alioth, Alcor, Mizar, Merak, Phecda and Megrez, which are all much younger than our Sun, were born together a few hundred million years ago within the same nebula. They are all between 78 and 84 light-years away.

2

ALCOR AND MIZAR
APPARENT DOUBLE STAR

DISTANCE: **81** LY FOR ALCOR AND **78** LY FOR MIZAR
DIAMETER: **5 MILLION KM**
LUMINOSITY: **60 SUNS**

The most famous star couple shines in the middle of the handle of the Big Dipper, between Alioth and Alkaid.

In ancient times, people used this mythical couple to test their vision, as stargazers still do today. Mizar and Alcor are indeed easily distinguishable with the naked eye under a clear country sky. When seen through binoculars, they form a very attractive pair. However, the two celestial siblings are separated from each other by nearly three light-years. Unlike true double stars, they are not physically connected by gravitational pull, even though they do move together through our galaxy.

3

MIZAR
DOUBLE STAR

DISTANCE: **78** LY
DIAMETER: **5 MILLION KM**
LUMINOSITY: **60 SUNS**

Mizar has another lovely surprise in store for experienced observers: it too is a bright double star — but a real one.

Alcor and Mizar form such a striking double star that it's easy to miss Mizar's true nature. But even through the smallest telescope you can see that it too has a small companion three times fainter than itself. The two stars, with their very pure white sparkle, orbit around each other at a distance of about 60 billion kilometers. But they are relatively close to one another compared to their partner Alcor, which is 500 times further away from them both. Whether true or false double stars, this apparent trio is a spectacular sight.

THE SUN AND THE STARS

Stars are gigantic spheres of gas, often measuring several million kilometers in diameter, that have surface temperatures of several thousand degrees. The Sun is an average star, typical of the galactic population. It measures 1.4 million kilometers in diameter and has a mass a thousand times greater than that of all of the planets in the Solar System combined. Solar gas, like the gas of most stars, is made up of approximately 70% hydrogen, 28% helium and 2% other atoms such as carbon, nitrogen and oxygen. The core of the Sun, with a temperature reaching 15 million degrees Celsius, is a nuclear fusion reactor. With each passing second, the Sun converts 4 million tons of matter into pure energy: 390 million billion billion watts, to be exact! The biggest stars are a hundred times more massive than the Sun, and shine over a million times brighter than it as well. The smallest stars are 20 times less massive and 1,000 times less brilliant.

See photo on right page

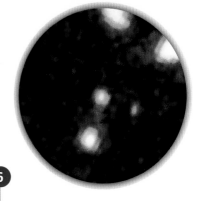

4 ALULA AUSTRALIS
DOUBLE STAR

DISTANCE: 26 LY
DIAMETER: 1.5 MILLION KM
LUMINOSITY: 1 SUN

Located south of the Big Dipper, this small double star has been fascinating observers for more than two centuries.

Alula Australis is one of the double stars closest to Earth. It is composed of two stars that are very similar, both in size and in brightness, to our star, the Sun. They are separated by only 3 billion kilometers and revolve around each other over a period of less than 60 years. For amateur astronomers, following this couple at regular intervals — best done with a 150-mm telescope — is a wonderful experience. In just a few years, the slow movement of the two companions can be perceived.

5 GALAXIES IN URSA MAJOR
SPIRAL GALAXIES M81 AND M82

DISTANCE: 12 MILLION LY
DIMENSIONS: 80,000 LY

These two spiral galaxies have been pirouetting around each other for several hundred million years and perform a cosmic *pas de deux*.

Although invisible to the naked eye, and beautifully distinct through binoculars, these two galaxies are truly wonderful when observed together through a telescope. M81, a giant galaxy populated by about a thousand billion stars, wraps thousands of young supergiant stars in its spectacular spiral arms. Its companion, M82, one of the brightest galaxies known in our small corner of the cosmos, is a spiral seen in profile. Its vast clouds of gas and dust hide its blazing core.

6 HDF4-473.0
PRIMORDIAL GALAXY

DISTANCE: 14 BILLION LY
DIMENSIONS: 10,000 LY

This galaxy, which is found at the edge of the known universe, allows astronomers to take a breathtaking plunge into the depths of time.

Discovered by the Hubble Space Telescope and studied by the largest telescope in the world, Galaxy HDF4-473.0 can only be viewed by giant instruments. And with good reason: to date, it is one of the most remote celestial bodies known. If the theory that the universe is 14 billion years old holds true, then the light from this galaxy has taken 13 billion years to reach us. We are therefore seeing HDF4-473.0 today as it was less than a billion years after the Big Bang. The very first stars of the universe shine from inside this primordial galaxy.

HERCULES

THE EARLY NIGHT SKY

Located between the small, shiny constellations of Corona Borealis and Lyra, this large, pale constellation includes no truly bright star. However, the outline of the mythological hero is easy to detect, stretching his limbs around a very characteristic trapezium of stars. Hercules killed Draco, the Dragon, which is always by its side in the sky. Today, astronomers are especially interested in Hercules' famous star cluster, and are searching for planets hidden by the light of its stars.

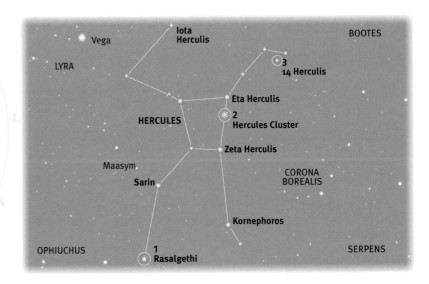

1

RASALGETHI
RED SUPERGIANT STAR

DISTANCE: **400** LY
DIAMETER: **600** MILLION KM
LUMINOSITY: **900** SUNS

 In an area fairly empty of stars at the south of the constellation, shines one of the most impressive supergiants in the galaxy.

When observed through binoculars or a telescope, the distant Rasalgethi clearly appears to be tinted red. This supergiant is truly an extraordinary size. If it were located in the center of the Solar System in the place of our Sun, it would incorporate the orbits of all the planets up to Mars. Its gaseous envelope is so delicate and fragile that it appears ready to burst out into space at a moment's notice. The temperature on the surface of its atmosphere varies cyclically between 2,000°C and 2,400°C.

2

HERCULES CLUSTER
GLOBULAR CLUSTER M13

DISTANCE: **24,000** LY
DIMENSIONS: **200** LY

 Between the stars Eta and Zeta Herculis lies a true giant: the great Hercules Cluster is the largest globular cluster in the Northern sky.

Although it is very distant, over 200 million billion kilometers away from us, M13 is visible to the naked eye on a very dark night. Through binoculars it looks like a small bright cloud. With a telescope it looks enchanting: the cloud transforms into a milky sphere studded with hundreds of faint stars. M13 contains over a million stars born some 10 billion years ago. Only the brightest among them, the red giants, are accessible through amateur instruments.

3

14 HERCULIS
STAR

DISTANCE: **60** LY
DIAMETER: **1** MILLION KM
LUMINOSITY: **0.75** SUN

 This moderately bright star, which is a bit smaller than the Sun, may be surrounded by an entire planetary system.

When observed through binoculars or a small telescope, 14 Herculis does not appear to have any distinguishing features. It has less mass than the Sun, and its life expectancy is just over 15 billion years. But astronomers have discovered a giant planet with a mass about three times that of Jupiter revolving around it every four and a half years, located 375 million kilometers away. Since its discovery, 14 Herculis has been constantly studied by observatories, to see if it may have other smaller planets in its orbit.

CARINA

THE EARLY NIGHT SKY

Traversed by the luminous meandering of the Milky Way, this vast constellation, which is easy to recognize from its location just next to Crux, the Southern Cross, is home to both the second brightest star in the sky, Canopus, and the most luminous nebula, Eta Carinae. Carina, the Keel, is a southern constellation that is very difficult to see from more northern latitudes. But it can be seen for a few weeks every year from North Africa, the Middle East and the southern United States.

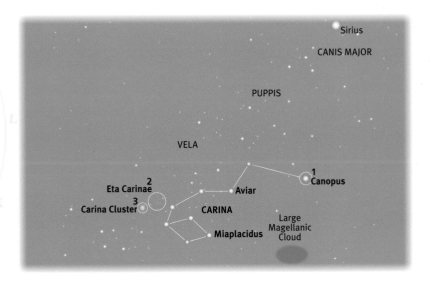

1

CANOPUS
SUPERGIANT STAR

DISTANCE: 300 LY
DIAMETER: 100 MILLION KM
LUMINOSITY: 14,000 SUNS

Canopus, found at the western point of Carina, is the only star in the entire sky that can rival Sirius' sparkle.

For the early Greeks, Canopus was the star that guided the troops of King Meneleas after the Trojan War. They could see it in the Mediterranean, drifting slowly just above the southern horizon, like a ship passing in the night far off from the shore. This supergiant has come to mid-life and is currently shining with a very pure white sparkle. However, in a few hundred thousand years, Canopus' envelope will become extremely dilated, and cool down. Then it will turn into a red supergiant before violently exploding as a supernova.

2

ETA CARINAE
NEBULA NGC 3372

DISTANCE: 8,000 LY
DIMENSIONS: 200 LY

More vast than the Orion Nebula, NGC 3372 harbors the brightest stars in the entire galaxy.

This is the most impressive nebula visible to the unaided eye. Through binoculars, NGC 3372 reveals its magnificent gas veils of hydrogen, helium and oxygen, which are heated by its brightest star, Eta, found in its center. Through a telescope, the sight of this interstellar chaos becomes truly spellbinding and almost surreal. The supergiant Eta is a genuine monster: it is among the most massive stars known, emits 5 million times more energy than the Sun and threatens to explode as a supernova any time now.

3

THE CARINA CLUSTER
STAR CLUSTER NGC 3532

DISTANCE: 1,500 LY
DIMENSIONS: 25 LY

NGC 3532, located quite close to the Eta Carinae nebula, stands out against the flurry of stars that surround it in the Milky Way. It is considered to be one of the most beautiful star clusters.

Although visible to the naked eye, NGC 3532 offers an even more beautiful sight through binoculars. But a modest wide-field telescope is best to truly appreciate it. This type of instrument will clearly reveal more than a hundred bright stars against the backdrop of the Milky Way. The cluster, which is less than 100 million years old, contains a large number of young, very white supergiants, distant twins of the star Canopus.

CENTAURUS

THE EARLY NIGHT SKY

This splendid constellation of the southern sky is well known as the starry jewel box containing Crux, the Southern Cross. Its two most luminous stars, Alpha Centauri and Beta Centauri, actually act as celestial signposts for the Cross. Teeming with remarkable stars, Centaurus, the Centaur, also houses Omega Centauri, a celestial object visible to the naked eye, which the ancients thought was a star. In reality, it is a remote, dense cluster of several million stars, arguably one of the most beautiful objects in the heavens

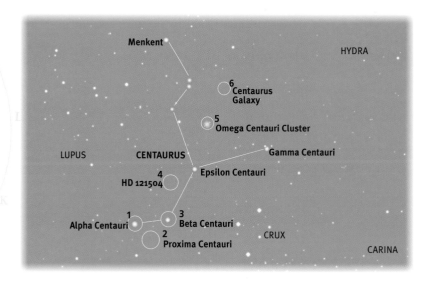

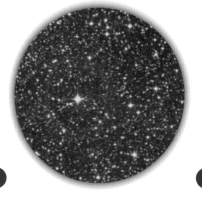

1

ALPHA CENTAURI
TRIPLE STAR

DISTANCE: **4.4** LY
DIAMETER: **1.5 MILLION KM**
LUMINOSITY: **1.5 SUNS**

This very famous star, the third brightest in the sky after Sirius and Canopus, has long been the preferred destination of science-fiction authors.

Alpha Centauri owes its fame to its proximity to Earth. At 4.4 light-years away from us, it is the nearest star visible to the naked eye. A telescope with a magnifying power of 100 can reveal what is hidden inside, namely a very close couple. Its companion, a small orange star, orbits around it every 80 years at a distance varying between 1.5 and 5 billion kilometers. However, only very large instruments can disclose that around this couple revolves a third, very faint star: Proxima Centauri.

2

PROXIMA CENTAURI
RED DWARF STAR

DISTANCE: **4.22** LY
DIAMETER: **200,000 KM**
LUMINOSITY: **0.00006 SUN**

A special note on this tiny red dwarf lost in the middle of the photograph: Proxima Centauri is the closest star to the Earth.

Very few astronomers have observed this dwarf star revolving around Alpha Centauri and its companion. Proxima Centauri shines 15,000 times less brightly than our Sun, making it invisible to the naked eye and through binoculars. It is only 4.22 light-years from Earth, that is 40,000 billion kilometers. In comparison, the Sun is a mere eight light-minutes away, that is 150 million kilometers, from Earth. Proxima is a red dwarf, a typical representative of the largest star population in the galaxy, which has perhaps several hundred billion of these unobtrusive stars.

3

BETA CENTAURI
BLUE SUPERGIANT STAR

DISTANCE: **500** LY
DIAMETER: **50 MILLION KM**
LUMINOSITY: **10,000 SUNS**

Beta Centauri and its "neighbor" Alpha, together designate the front hooves of the mythological Centaur.

To the naked eye, Beta Centauri forms, with Alpha Centauri, a handsome celestial couple clearly visible from tropical and southern latitudes. When we see these two stars shining side by side, we imagine that they must be very close to one another. Not at all: Beta is over a hundred times further from the Sun than Alpha. In terms of their apparent brightness in the celestial dome, this star is in eleventh place, but it is actually a supergiant, shining 6,000 times more brightly than Alpha Centauri and 10,000 times more brightly than the Sun.

THE BIG BANG THEORY

We know today that the universe has not always existed. In the 1920s, following the work of Albert Einstein, astronomers discovered that the universe was expanding: on an immense scale, all galaxies are moving away from each other, and the further away they are, the faster they move. Then, in the 1960s, it was discovered that a low-level radiation was filling all of space. It was baptized fossil radiation and was attributed to a previous hot and dense state of the universe. Finally, more recently, work on the age of star clusters such as Omega Centauri made it possible to accurately date the age of the oldest stars: a little under 15 billion years old. Today, all indications confirm this thesis, and cosmologists are sure that, about 14 billion years ago, the universe was created out of a sort of primordial explosion, the Big Bang. The galaxies and the stars were formed several million years after the Big Bang, when the fluid of pure energy that was generated in its quick expansion gradually cooled.

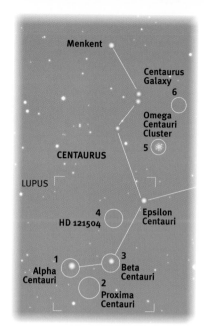

4

5

6

HD 121504
STAR

DISTANCE: **144** LY
DIAMETER: **1.4** MILLION KM
LUMINOSITY: **1** SUN

Halfway between Beta and Epsilon Centauri, there is an unpretentious star, shown in the middle of the photograph, that hides in its flames a planet almost as big as our Jupiter.

This distant twin of our Sun is invisible to the naked eye but easy to find through binoculars. Astronomers directing their telescopes at it all feel a twinge of excitement: they know that a planet almost as massive as Jupiter orbits around it. This planet is probably a real furnace. It revolves in just over two months, at only 50 million kilometers from its star. The surface or atmospheric temperature is probably higher than 600°C.

OMEGA CENTAURI CLUSTER
STAR CLUSTER NGC 5139

DISTANCE: **18,000** LY
DIMENSIONS: **300** LY
LUMINOSITY: **1** MILLION SUNS

Omega Centauri is simply described: it is the densest, the biggest and the most beautiful of all the star clusters in our galaxy.

Despite its distance, Omega Centauri is a clearly perceived naked-eye object. It resembles a small, faint and hazy star. But to be appreciated in all of its beauty, this cluster needs to be observed through a small telescope. The sight is enchanting. Hundreds of stars appear, seemingly stuck together. In a large telescope, they become countless. In reality, Omega Centauri contains over 10 million stars, some of which are only a few light-weeks away from one another.

CENTAURUS GALAXY
GALAXY NGC 5128

DISTANCE: **12** MILLION LY
DIMENSIONS: **100,000** LY

This galaxy, which is among the closest to ours, is a strange celestial object, probably the remnant of the merger of two giant galaxies.

Although it can easily be seen through binoculars, the Centaurus Galaxy offers a spectacular image through a small amateur telescope. This giant elliptical galaxy is traversed by a dark wide band composed of gas and dust. This is all that remains of a doomed spiral galaxy that was literally swallowed up in the distant past by an elliptical galaxy. The Centaurus Galaxy probably contains over a thousand billion stars, and astronomers have found a giant black hole at its core.

THE OMEGA CENTAURI CLUSTER

*T*his *photograph portrays only Omega Centauri's brightest stars; in reality the cluster contains over 10 million stars, all a few billion years old. The field in this image is about 100 light-years. In the center of the cluster, the stellar density is 25,000 times greater than it is in the solar environment.*

CRUX

THE EARLY NIGHT SKY

This is the smallest but also one of the most famous and most beautiful of the constellations. Seen only from the Southern latitudes, Crux, the Southern Cross, appears slowly above the horizon, due south, at the end of winter. The sight is splendid even without optical enhancement. Right in the middle of the Milky Way, among the nearby Centaurus bodies, the sparkle of the four bright stars that form Crux appears to be heightened by the nearby presence of a strange, dark zone empty of stars, called the Coal Sack.

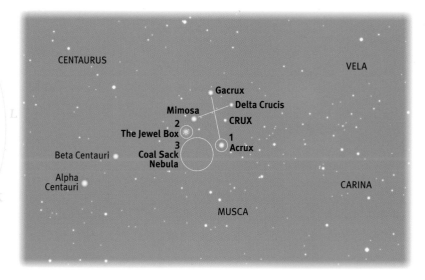

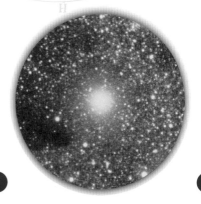

ACRUX
DOUBLE STAR

DISTANCE: 300 LY
DIAMETER: 30 MILLION KM
LUMINOSITY: 3,000 SUNS

Marking the foot of Crux, Acrux ranks 13th among the most luminous stars in the sky.

To the naked eye, Acrux, the brightest of the constellation's stars, appears to be nestled against the Coal Sack Nebula. Acrux is actually a double star composed of two blue supergiants, both visible through small amateur optical instruments. The brightest of the two has a luminosity of 3,000 suns and its companion shines like 2,000 suns. Its companion orbits at about 80 billion kilometers from it, but its period of revolution is still unknown. This interval is probably measured in centuries, or even in millennia.

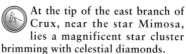

THE JEWEL BOX
STAR CLUSTER NGC 4755

DISTANCE: 7,600 LY
DIMENSIONS: 50 LY

At the tip of the east branch of Crux, near the star Mimosa, lies a magnificent star cluster brimming with celestial diamonds.

Although perceptible to the naked eye, this cluster is particularly spectacular when seen through binoculars. Then it appears as a small silvery cloud, scattered with stars and bordered by the dark, dense clouds of the Coal Sack Nebula. The Jewel Box contains several dozen blue supergiants similar to Deneb and Rigel, each shining like 100,000 suns. A red supergiant, which can be seen almost in the center, has a sparkle comparable to that of Betelgeuse. NGC 4755, one of the youngest clusters known in the galaxy, was formed less than 10 million years ago.

COAL SACK NEBULA
DARK NEBULA

DISTANCE: 600 LY
DIMENSIONS: 60 LY

The Coal Sack Nebula appears as a dark hole quite close to Acrux, silhouetted against the tumultuous background of the Milky Way.

While perfectly visible to the naked eye, the Coal Sack Nebula should be observed through binoculars in order to see it in striking contrast to the bright stars in Crux. This is a dark nebula, measuring about 60 light-years, and is about 600 light-years away. It is made up of cold gases and interstellar dust that completely absorb the light of its background stars. The dark, cold globules of hydrogen and helium within it are slowly condensing and will one day give birth to new stars.

LYRA

THE MIDNIGHT SKY

This small but charming constellation represents Orpheus' lyre in the sky. Its main star, Vega, meaning "eagle" in Arabic, is one of the brightest stars in the heavens. Just below Vega, the stars of Lyra form an almost perfect diamond shape, in which one of the most beautiful nebulas, the Ring Nebula, is concealed. Deneb in Cygnus, Altair in Aquila and Vega in Lyra together make up the great Summer Triangle in the northern sky, through which the Milky Way passes.

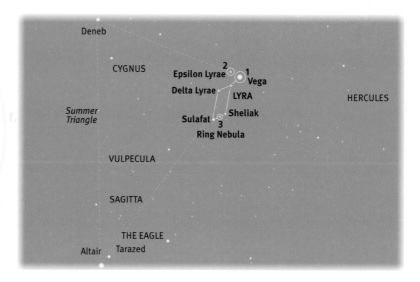

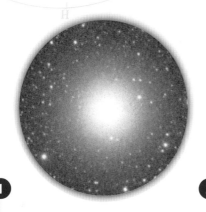

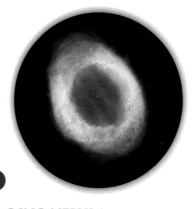

VEGA
STAR

DISTANCE: **25.3** LY
DIAMETER: **4** MILLION KM
LUMINOSITY: **50** SUNS

Vega is the fifth brightest star in the sky, after Sirius, Canopus, Alpha Centauri and Arcturus. It also plays the starring role in the Summer Triangle, which it forms with Deneb and Altair.

To the naked eye, Vega's brilliance practically eclipses the four small stars that make up Lyra's diamond. But in fact it stands out only because it is relatively close to Earth. After all, Vega is "only" 50 times brighter than the Sun. When observed through binoculars or a telescope, this young star seems to have beautiful bluish glints. Its surface temperature reaches 9,500°C, almost twice that of the Sun.

EPSILON LYRAE
MULTIPLE STAR

DISTANCE: **130** LY
DIAMETER: **3** MILLION KM
LUMINOSITY: **20** SUNS

Just next to Vega, one of the most impressive star systems in the sky holds special surprises for its admirers.

To the unaided eye, Epsilon Lyrae appears as a single, fairly distinct star. With a pair of binoculars alone it becomes clear that it has two components offering, with nearby Vega, a magnificent sight. However, an amateur telescope with a magnifying power of 150 to 300 will reveal that Epsilon Lyrae actually contains four stars. Each component of the couple is itself a double. The brightest of the four stars is 20 times brighter than the Sun. The two couples are separated by about two light-months.

RING NEBULA
NEBULA M57

DISTANCE: **2,000** LY
DIMENSIONS: **1** LY

Nestled between the stars Sheliak and Sulafat, Nebula M57 looks strikingly like a small smoke ring.

Although invisible to the naked eye and through binoculars, the Ring Nebula is lovely to look at under a very clear sky through a 100- to 300-mm telescope with a magnifying power of 100 to 200. This shell of gas was expelled a few thousand years ago by a red giant star in the process of dying out. Having blown most of its envelope into space, the star became a tiny white dwarf. It is now slowly fading and lights up the nebula, which will dwindle over time and disappear in less than 10 thousand years.

CYGNUS

THE MIDNIGHT SKY

One of the prettiest constellations, Cygnus dominates the northern summer sky. Its almost perfect cross shape symbolizes the Swan, one of the many animal incarnations of Zeus. This long-necked Swan flying south in the company of the Eagle takes the starry high road of the Milky Way. It contains a wealth of bright, distant stars, such as the extraordinary and unique Deneb, an interstellar beacon that is so powerful it lights up hundreds of light-years of space around it.

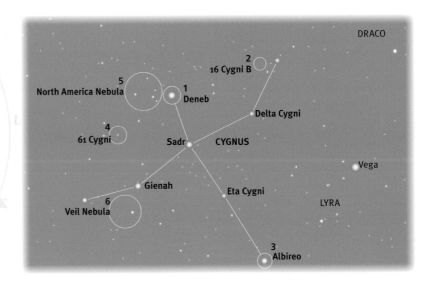

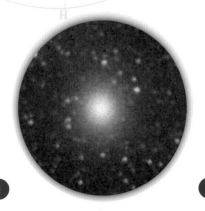

DENEB
BLUE SUPERGIANT STAR

DISTANCE: **3,000** LY
DIAMETER: **280 MILLION KM**
LUMINOSITY: **300,000 SUNS**

Deneb shines at the tail of the Swan. But this supergiant also marks one of the three points of the famous Summer Triangle, joined by its neighbors Vega in Lyra and Altair in Aquila.

Deneb is one of the brightest stars known in the galaxy and one of the most remote visible to the naked eye. Its mass is probably in excess of 30 suns. Born about a million years ago, this blue supergiant, almost 200 times bigger and 300,000 times brighter than the Sun, burns its nuclear fuel at a dizzying rate. Its life will end in 2 or 3 million years in the gigantic fireworks display of a supernova explosion.

16 CYGNI B
STAR

DISTANCE: **70** LY
DIAMETER: **1.5 MILLION KM**
LUMINOSITY: **1 SUN**

A planet slightly bigger than Jupiter revolves around this small star, which is a true twin of the Sun.

16 Cygni B is quite easy to find through binoculars or a small telescope, sitting right next to its neighbor, 16 Cygni A. But observing it through a telescope may mean that you are actually admiring a distant solar system. This star's size and brightness greatly resemble our Sun's. It also has at least one planet with a mass just a bit greater than Jupiter's. This invisible body, discovered thanks to the gravitational pull it exerts on its star, revolves in just over two years, at 250 million kilometers from the star on a very elliptical orbit.

ALBIREO
DOUBLE STAR

DISTANCE: **400** LY
DIAMETER: **100 MILLION KM**
LUMINOSITY: **700 SUNS**

Albireo marks the beak of the Swan. It is actually a star couple presenting a feast of contrasting colors.

Although visible through binoculars, a small telescope is needed to truly do justice to this magnificent binary star. It is comprised of an orange-yellow giant star 700 times brighter than the Sun, and a small blue star 100 times brighter than the Sun. It is easy to see the color of the two stars even through the smallest amateur instruments. Although the pair have been studied in great detail since 1832, no movement has yet been recorded. It is still not known if Albireo is a true double star or just an apparent couple.

IN SEARCH OF OTHER PLANETS

This is one of the greatest scientific issues of the 21st century. Thanks to the 1995 discovery of the very first exoplanet, around the star 51 Pegasi, we now know that an almost infinite number of stars are accompanied by a procession of planets, some of which are similar to those in our own Solar System. In 2004 astronomers know of over 150 exoplanets surrounding stars from 10 to 200 light-years away. Some exoplanets revolve around stars that are true twins of our Sun. But it has also been found that some planets orbit around red dwarf stars and even double stars. We cannot yet photograph these remote planets. Their existence is deduced only from their gravitational effect on their star, or sometimes by the eclipses they cause. To date, only planets with a mass comparable to Jupiter's and Saturn's have been discovered; however, in the coming years astronomers will be searching among the stars for other planets with a mass comparable to Earth's.

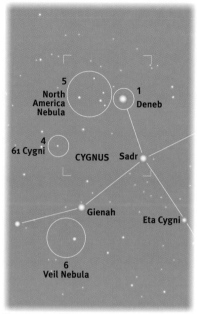

See photo on right page

61 CYGNI
DOUBLE STAR

DISTANCE: **11.4** LY
DIAMETER: **700,000** KM
LUMINOSITY: **0.06** SUN

Observers are guaranteed a very rare treat as they gaze upon this strange couple composed of two tiny red stars.

61 Cygni is one of the double stars located closest to Earth, just 110,000 billion kilometers away. It is visible to the naked eye on a clear summer night. Through binoculars, the partners are easily detectable, and through a telescope the sight is quite astonishing: the two stars are clearly quite orange in color. These are, in fact, two red dwarfs, 15 times and 25 times fainter than the Sun. It takes almost seven centuries for the smaller of the two to revolve around the larger one. The two stars are located about 12 billion kilometers from each other.

NORTH AMERICA NEBULA
NEBULA NGC 7000

DISTANCE: **3,000** LY
DIMENSIONS: **50** LY

Just a short distance from the supergiant Deneb is a vast cloud of gas with a very recognizable shape.

This large but unobtrusive nebula, lit up by Deneb, is best seen by the naked eye on a very clear night free of any light pollution. When observed through binoculars, the North America Nebula clearly shows the shape of its namesake. Its paleness contrasts markedly with Deneb, a true jewel of light against the velvety black sky. This nebula is very easy to photograph even with a fairly simple camera. Using extremely light-sensitive film, you only need about 30 seconds of exposure to capture its subtle pink and red-tinted wreaths of hydrogen.

VEIL NEBULA
NEBULAE NGC 6960-6992

DISTANCE: **1,500** LY
DIMENSIONS: **80** LY

This vast group of nebulas is the remains of a very ancient supernova that exploded some 5,000 years ago.

Nebulas are only visible on clear, transparent nights. Although it is visible through binoculars, the Veil Nebula (indexed by specialists under the numbers NGC 6960 and NGC 6992) is an especially wonderful sight through a large telescope. On photographs taken with recreational telescopes, it draws a vast, hazy loop that forms a bright halo around 52 Cygni, a star visible without optical aids. Its delicate veils were formed by the explosion of a big star and they are now spreading out through space at a speed of 60 kilometers per second.

AQUILA

THE MIDNIGHT SKY

Aquila, the Eagle, which at one time carried Zeus' lightning, flies majestically in one of the richest star fields in the sky. To the naked eye, the bright stars Altair, Alshain and Tarazed (Arabic and Persian names for the mythological bird) stand out against a galactic arm that follows the Great Rift, an immense dark zone that appears to divide the Milky Way into two. Another object of interest is a bright stellar cloud in Scutum, the Shield, that appears on the outer edges of Aquila and Sagittarius.

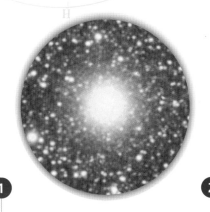

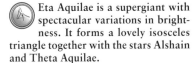

① ALTAIR
STAR

DISTANCE: **16.8 LY**
DIAMETER: **2 MILLION KM**
LUMINOSITY: **11 SUNS**

After Vega in Lyra and Deneb in Cygnus, Altair in Aquila is considered the "third beauty" in the celebrated Summer Triangle.

Altair is easily visible to the naked eye in the summer sky, where it shines with a lovely white sparkle. It is one of the stars closest to Earth, along with Sirius in Canis Major and Procyon in Canis Minor, which are located 8.6 light-years and 11.4 light-years away, respectively. Its light takes 16.8 years to cover the 160,000 billion kilometers that separate it from us. It is overwhelming to think that the American space probe Pioneer 11, which left Earth in 1973 and is moving toward the constellation Aquila, will pass Altair in about a million years.

② ETA AQUILAE
RED SUPERGIANT STAR

DISTANCE: **1,200 LY**
DIAMETER: **110 MILLION KM**
LUMINOSITY: **4,000 SUNS**

Eta Aquilae is a supergiant with spectacular variations in brightness. It forms a lovely isosceles triangle together with the stars Alshain and Theta Aquilae.

Eta Aquilae is a cepheid star, that is to say, a red supergiant whose enormous atmosphere varies cyclically in volume, temperature and luminosity. The star's intrinsic sparkle increases by a factor of two over a period of seven days and four hours. During this time it shines like 4,000 suns, twice as brightly as its usual 2,000 suns. Its variations in brightness are visible to the naked eye with the help of nearby Alshain. When Eta Aquilae is at its maximum, the two stars have exactly the same brightness.

③ WILD DUCK CLUSTER
STAR CLUSTER M11

DISTANCE: **5,600 LY**
DIMENSIONS: **15 LY**

Just beneath the constellation Aquila, this splendid cluster dominates the constellation Scutum, a vast star field with exceptional brilliance.

M11 does not actually belong to Aquila but rather to the small nearby constellation, Scutum. This cluster, although visible through binoculars, can be seen to best effect using a small refractor. Then its magnificent swarm of several dozen stars becomes evident. Overall, the Wild Duck Cluster contains over 1,000 stars, which are located very close to each other, given that they are separated on average by less than one light-year. These stars were born at the same time about 500 million years ago.

SCORPIUS

THE MIDNIGHT SKY

This is one of the most impressive constellations in the sky. The huge arthropod, which, according to mythology, slew Orion, is remarkably well drawn in the sky. Its eye is marked by the brilliant Antares, its claws by Graffias and Pi Scorpii, its immense body by a number of bright stars and its stinger, at the back, by Shaula, "she who stings." Located, like its neighbor Sagittarius, right in the heart of the Milky Way, this constellation offers a wonderful galactic panorama.

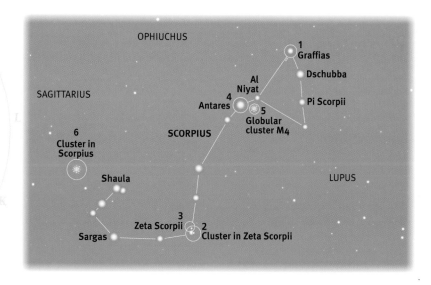

OPHIUCHUS

1 Graffias

Dschubba

Al Niyat

4 Antares

Pi Scorpii

5 Globular cluster M4

SAGITTARIUS

SCORPIUS

6 Cluster in Scorpius

LUPUS

Shaula

3 Zeta Scorpii

2 Cluster in Zeta Scorpii

Sargas

GRAFFIAS
DOUBLE STAR

DISTANCE: 500 LY
DIAMETER: 20 MILLION KM
LUMINOSITY: 2,000 SUNS

 Graffias is at the tip of the constellation and represents one of the mythological creature's claws.

Graffias is clearly visible with the naked eye, but its true nature can only be observed through a telescope. It is a blue giant side by side with a small star about 10 times fainter than it and also bluish in color. Astronomers have been monitoring the couple for nearly two centuries, yet there has been no noticeable change. The space between the two bodies remains stable and they do not appear to revolve around each other. Perhaps the two stars are not connected by gravity, but are a false couple that appear to be close together in the sky due to a sight-line illusion.

CLUSTER IN ZETA SCORPII
STAR CLUSTER NGC 6231

DISTANCE: 6,000 LY
DIMENSIONS: 10 LY

This formidable agglomeration of supergiants contains Zeta Scorpii, one of the brightest stars in the galaxy, located just on its southern edge.

The star cluster NGC 6231, one of the youngest in the galaxy, can only be seen with the naked eye in the clear skies of the tropics, where it appears like a tiny hazy spot dominated by the sparkle of Zeta Scorpii. A telescope reveals some 30 bright stars, sparkling like diamonds. All of these stars are blue supergiants trying to outshine Rigel in Orion or Deneb in Cygnus, and they were all born about one or two million years ago.

ZETA SCORPII
BLUE SUPERGIANT STAR

DISTANCE: 6,000 LY
DIAMETER: 200 MILLION KM
LUMINOSITY: 300,000 SUNS

Shining at the Scorpion's tail, Zeta Scorpii is the brightest star in the cluster NGC 6231.

Although Zeta Scorpii is the most distant star visible to the naked eye, it can only truly be observed from the tropics. Comparable to the beautiful Deneb in Cygnus in its mass, size and luminosity, this blue supergiant is twice as far away and therefore appears much fainter. Like most other supergiants, Zeta Scorpii is permanently discharging part of its mass in the form of violent puffs of gas with a temperature of several million degrees Celsius. A dramatic supernova explosion will occur in one or two million years.

HOW OLD ARE THE STARS?

What is it that allows the stars to endure? To continue that existence they must maintain a delicate balance between nuclear force, which tends to make them explode, and gravitational force, which tends to make them implode. This is how a star like the Sun burns the hydrogen in its core with clockwork regularity and has maintained its size, temperature and brightness over nearly 5 billion years with such impressive stability. The Sun will keep up this balance of forces for about another 4 billion years, before turning into a very unstable red giant star, and finally into a white dwarf star that will slowly fade away. The smallest stars are the ones with the greatest longevity. The oldest known stars have been around for about 14 billion years. Some of them have a life expectancy of 100 billion years. On the other hand, the most massive stars, such as Antares or Zeta Scorpii, quickly burn away their young lives. Supergiants burn out in just a few million years before violently exploding as supernovas.

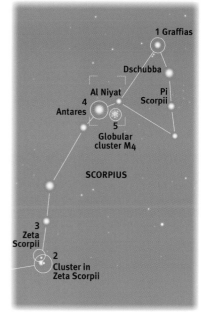

See photo on right page

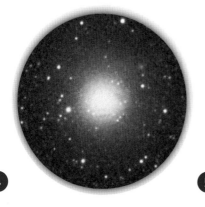

4

5

6

ANTARES
RED SUPERGIANT STAR

DISTANCE: 700 LY
DIAMETER: 1 BILLION KM
LUMINOSITY: 10,000 SUNS

Antares owes its name — which means "rival of Ares," the ancient Greek god of war — to its fiery color. The red star shines at the very heart of Scorpius.

Antares is the brightest and biggest of the red supergiants visible in our region of the galaxy. Its red sparkle is further intensified through binoculars. The surface temperature of this immense sphere a billion kilometers in diameter is nearly 2,800°C. A telescope reveals that there is a tiny greenish companion nestled against it, which is drowned out by its bright light. This companion, 100 times fainter than Antares, takes nearly a thousand years to revolve around it at a distance of over 80 billion kilometers away.

GLOBULAR CLUSTER M4
STAR CLUSTER

DISTANCE: 7,000 LY
DIMENSIONS: 50 LY

This is one of the oldest star clusters known in the galaxy, almost hidden away just beside Antares.

Cluster M4 is very easy to see through binoculars, but looks like a large pincushion of faintly shining stars when seen through a telescope. In reality, it contains about 500,000 stars crowded very close together. Like the 150 globular clusters known in the galaxy, M4 is a dense gathering of stars that is almost perfectly spherical. It contains only very old stars, all born together about 12 billion years ago.

CLUSTER IN SCORPIUS
STAR CLUSTER M7

DISTANCE: 800 LY
DIMENSIONS: 40 LY

This vast and splendid star cluster, marked by the star Shaula, seems to be an extension of the Scorpion's stinger.

If you have the opportunity to gaze at cluster M7 with the naked eye — you have to be in the southern part of the U.S., Mexico or farther south — you will see in it a bright cloud superimposed on a background strewn with stars of the Milky Way. With binoculars, some 15 stars stand out against this background, and with a telescope a few dozen more appear. M7 is a young cluster. Its stars were all born together inside the same nebula about 250 million years ago.

SAGITTARIUS

THE MIDNIGHT SKY

Being able to admire the Milky Way in the constellation Sagittarius, where the mysterious center of the galaxy is hidden, beneath a crystal clear tropical sky is a wonderful and unique experience. This section of the sky is so rich in distant clusters and nebulas and the Milky Way shines so intensely at this point, that the stars closest to Sagittarius literally appear to fade. This galactic panorama, which is visible without optical assistance, is truly breathtaking.

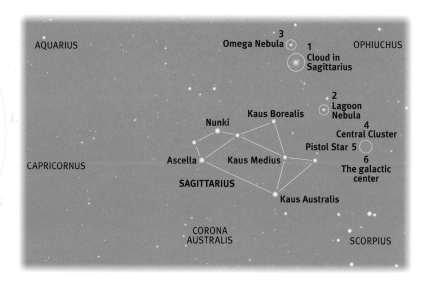

AQUARIUS
3 Omega Nebula
1 Cloud in Sagittarius
OPHIUCHUS
2 Lagoon Nebula
Kaus Borealis
Nunki
4 Central Cluster
Pistol Star 5
6 The galactic center
CAPRICORNUS
Ascella   Kaus Medius
SAGITTARIUS
Kaus Australis
CORONA AUSTRALIS
SCORPIUS

1

CLOUD IN SAGITTARIUS
STAR CLUSTER M24

DISTANCE: **16,000** LY
DIMENSIONS: **500** LY

The cloud in Sagittarius lies to the northwest of the constellation, right inside the Milky Way, in an area brimming with star clusters.

Although the Milky Way is especially impressive when observed at the zenith, from the Andes for example, observers in the northern hemisphere can also catch a glimpse of its main star clusters, which can be seen by the naked eye. The brightest among them, M24 in Sagittarius, is very easy to spot, even when observed just above the horizon. The sight is spectacular through a pair of powerful binoculars, which will reveal a shimmering of stars, bordered by dark areas and surrounded by other stellar clusters and bright nebulas.

2

LAGOON NEBULA
NEBULA M8

DISTANCE: **5,200** LY
DIMENSIONS: **60** LY

Just south of the cloud in Sagittarius lies one of the most beautiful nebulas in our region of the galaxy.

The Lagoon is one of the very rare nebulas visible to the unaided eye. It is easy to see under a very clear sky, but it can only truly be appreciated through a small amateur instrument with a magnifying power of 50 to 100. The sight promises to be striking, with myriad bright stars standing out against a vast cloud of gas in a surreal light. M8 is scattered with milky condensation and is bisected into two lobes by a dark channel, which gives it the name Lagoon Nebula.

3

OMEGA NEBULA
NEBULA M17

DISTANCE: **5,000** LY
DIMENSIONS: **30** LY

Situated just at the border o the constellation Scutum, nort of the cloud in Sagittarius, th Omega Nebula is a favorite celestia object of amateur astronomers.

Despite its remoteness, this very brigh and well-defined nebula is remark ably clear through binoculars. Whe observed through a small telescop it appears to have a presence tha few other nebulas can claim. Th distinct shape of its brilliant gaseou veils of hydrogen gives Nebula M1 the majesty of a sphinx—a sphin that presides over more than 3 light-years and glides through spac as if in a dream.

THE MILKY WAY, OUR GALAXY

It is very difficult to form an accurate image of our galaxy, because we are immersed deep within it, and it is impossible for us to look at it from the outside. However, on a clear summer night, beneath the tropical sky, we are able to make out the profile of a spiral galaxy strewn with dark clouds in the great silver band of the Milky Way. We now know that the galaxy is a large disk, thicker in the center and measuring about 100,000 light-years in diameter. The Sun is 26,000 light-years from the galactic center, where a black hole lurks, hidden behind thousands of light-years of gas and interstellar dust. The black hole has a mass about a million times greater than the Sun's, but is compressed into a sphere of less than 5 million kilometers in diameter. Nobody knows how many stars are slowly revolving in the Milky Way, but there are probably more than a trillion.

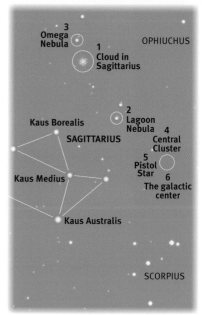

See photo on right page

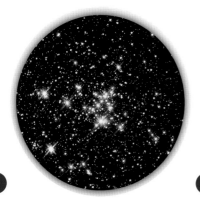

CENTRAL CLUSTER
SUPERGIANT STAR CLUSTER

DISTANCE: **26,000** LY
DIMENSION: **10** LY

Only 100 light-years from the very heart of our galaxy lies the Central Cluster, the most enormous group of stars known in the universe.

The Central Cluster can only be seen through infrared telescopes capable of piercing through the fog of gas and dust that obscures the central region of the galaxy. It contains several thousand stars less than 4 million years old. Several hundred of these stars are the brightest supergiants in the Milky Way and are ready to explode as supernovas at any moment. But these explosions will never be visible to the naked eye because their light will be completely absorbed by interstellar dust.

PISTOL STAR
BLUE SUPERGIANT STAR

DISTANCE: **26,000** LY
DIAMETER: **180 MILLION KM**
LUMINOSITY: **5 MILLION SUNS**

The Pistol Star, located within the Central Cluster, is the brightest star known in the universe.

This supergiant, an enormous celestial furnace, consumes as much energy in six seconds as the Sun does in one year. However, it is invisible through an optical telescope. Only infrared instruments are capable of observing it through the thick screen of dust that lies between it and us. The Pistol Star was born just over a million years ago. If it were placed in the center of the Solar System, it would douse the Earth in its flames. It is not surprising that the maelstrom of gas and dust turning at great speed around the galactic center sometimes gives birth to such colossal objects.

THE GALACTIC CENTER
GIANT BLACK HOLE

DISTANCE: **26,000** LY
DIAMETER: **5 MILLION KM**

At the very heart of our galaxy is a giant black hole, the bottomless abyss where stars are lost.

The thousands of billions of stars in the galaxy rotate around one single point in space and time called Sag A*. Only radio telescopes can demonstrate that this is a disk of matter that turns at an astounding speed around a black hole measuring 5 million kilometers in diameter, and imprisoning a mass equal to 2 million suns in its dark recesses. One day, future giant telescopes will show us how this abyss swallows up the stars that pass within its reach.

PEGASUS

THE LATE NIGHT SKY

Although this immense constellation does not contain any truly bright stars, it is still easy to recognize by the end of summer. Its central region sketches out an almost perfect square. Pegasus was the mythological winged horse born of the blood of Medusa, one of the gorgons killed by Perseus. Markab, the name of its main star, means "shoulder," while Scheat, its neighbor, means "hock," or "back of the knee." East of Pegasus lies Andromeda, the daughter of King Cepheus, who Perseus rescued from the clutches of the monster Cetus.

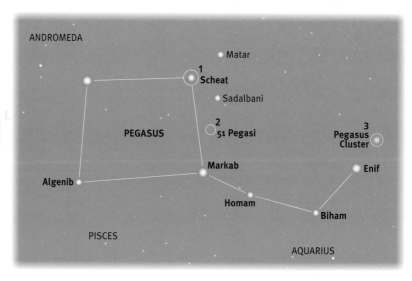

ANDROMEDA
Matar
1 Scheat
Sadalbani
2 51 Pegasi
PEGASUS
3 Pegasus Cluster
Enif
Markab
Algenib
Homam
Biham
PISCES
AQUARIUS

1 SCHEAT
RED SUPERGIANT STAR

DISTANCE: 200 LY
DIAMETER: 200 MILLION KM
LUMINOSITY: 300 SUNS

Scheat marks the northwest side of the Square of Pegasus. If placed at the center of the Solar System, this gigantic star would swallow up Mercury, Venus and even Earth.

Scheat, a beautiful red supergiant, only shows its color when examined through a pair of binoculars or a small telescope. However, even a powerful recreational telescope only portrays it as a pinprick of light with reddish reflections. A special instrument called an interferometer is required to accurately assess its huge size. Scheat is a pulsating star, whose gaseous envelope, which is 140 times larger than our Sun's, has a slow irregular heartbeat.

2 51 PEGASI
STAR

DISTANCE: 50 LY
DIAMETER: 1.5 MILLION KM
LUMINOSITY: 1 SUN

Not far from Scheat, west of the Square of Pegasus, lies a modest star that has now become famous: it was around this star that the first exoplanet in history was discovered.

51 Pegasi, a true twin of the Sun, is located at the extreme limit of visibility of the human eye, and can only be seen through binoculars and, of course, a telescope. In 1995, it made a dramatic entrance on the astronomical stage when a strange planet was discovered revolving around it once every four days. Although separated from Scheat by some 7.5 million kilometers, the planet is constantly whipped by the star's fires. Its burning atmosphere reaches over 900°C.

3 PEGASUS CLUSTER
STAR CLUSTER M15

DISTANCE: 40,000 LY
DIMENSIONS: 130 LY

This spectacular globular cluster is visible in the southwest of the constellation, just a short distance away from the bright star Enif.

On a clear autumn night, M15 is visible to the naked eye despite its great distance. Through binoculars it looks like a small, bright and hazy spot, but through a telescope the sight is truly spectacular. A powerful instrument clearly reveals hundreds of stars crammed together against a milky background. These stars, visible through an amateur instrument, are all supergiants. The cluster contains over a hundred thousand stars, all about 12 billion years old.

ANDROMEDA

THE LATE NIGHT SKY

This big constellation is very easy to see, stretching out toward the east from the Square of Pegasus not far from the Milky Way. Andromeda, visible starting at the end of summer, holds some of the marvels of the sky. However, when the beautiful Andromeda, daughter of Cepheus and Cassiopeia, appears on the horizon all astronomers' eyes and telescopes turn toward its famous spiral galaxy, which has been known since ancient times.

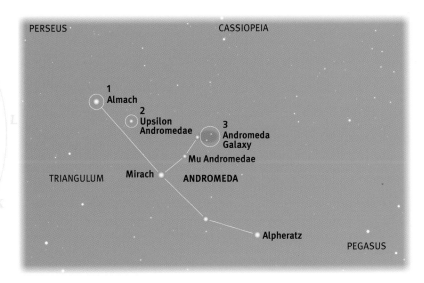

PERSEUS · CASSIOPEIA
1 Almach
2 Upsilon Andromedae
3 Andromeda Galaxy
Mu Andromedae
TRIANGULUM · Mirach · ANDROMEDA
Alpheratz
PEGASUS

1

2

3

ALMACH
MULTIPLE STAR

DISTANCE: **400** LY
DIAMETER: **140** MILLION KM
LUMINOSITY: **1,000** SUNS

Almach's name comes from the Arabic meaning "desert lynx." It is one of the most beautiful multiple stars in the sky.

Although clearly visible with the naked eye, the true beauty of Almach is revealed through a small amateur telescope. Even at a low magnifying power of 50 to 100, the experience will certainly be an unforgettable one: the star appears to be doubled and very colorful. Next to this bright red giant, a hundred times bigger and a thousand times brighter than the Sun, shines a small, young bluish object. The real surprise is that it is in fact a triple star. However, this can only be seen through very large telescopes.

UPSILON ANDROMEDAE
STAR

DISTANCE: **44** LY
DIAMETER: **2** MILLION KM
LUMINOSITY: **2** SUNS

One of the greatest astronomical discoveries made at the end of the 20th century was the finding that Upsilon Andromedae is surrounded by an actual planetary system.

When examined with the naked eye, binoculars or even a telescope, Upsilon Andromedae shows no trace of the planets orbiting around it. However, its planets are really there, drowned out by the overpowering light of their star. To date, astronomers know of three, with a mass from between one to four times greater than Jupiter. And the three-billion-year-old Upsilon Andromedae is probably also hiding other undiscovered planets in its flames.

ANDROMEDA GALAXY
SPIRAL GALAXY M31

DISTANCE: **2.8** MILLION LY
DIMSENSIONS: **200,000** LY

This very beautiful spiral galaxy, located just above the stars Mirach and Mu Andromedae, is a mirror image of our own Milky Way.

The Andromeda Galaxy is the most distant celestial body available to the naked eye. To admire it, you will need a very clear sky without the haze of city lights. While it is an enchanting sight through binoculars, examining it through a wide-field telescope is most exciting. You will see a vast silvery disk that creates a halo around a more luminous bulb, and two small satellite galaxies, M32 and NGC 205. M31 is an impressive spiral galaxy just like our Milky Way and probably contains more than a trillion stars.

THE ANDROMEDA GALAXY

The famous Andromeda Galaxy greatly resembles our own in size, mass and shape. The yellow bulb at the center is populated by old stars. In addition to its tens of billions of stars each comparable to the Sun, the disk is home to a few million blue supergiants that give it its characteristic color.

CASSIOPEIA

THE LATE NIGHT SKY

Cassiopeia is one of the most famous constellations because it is very easy to recognize. It draws a distinct W that revolves slowly around the North Pole and never goes below the horizon. Cassiopeia, the flamboyant mythical queen, sits enthroned in the sky beside her retiring spouse, Cepheus, and her daughter, the beautiful Andromeda. The diaphanous veils of the Milky Way adorn her with sumptuous, starry sapphires and diamonds.

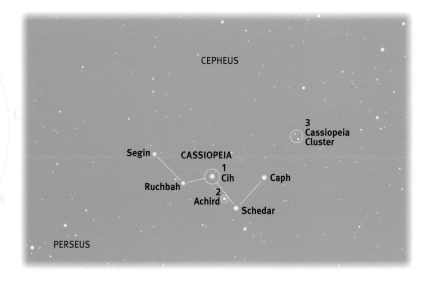

CEPHEUS

3 Cassiopeia Cluster

Segin

CASSIOPEIA

1 Cih

Caph

Ruchbah

2 Achird

Schedar

PERSEUS

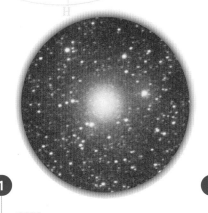

1

CIH
BLUE SUPERGIANT STAR

DISTANCE: **700 LY**
DIAMETER: **15 MILLION KM**
LUMINOSITY: **4,000 SUNS**

This distant star lies right in the middle of the W in Cassiopeia and surprisingly changes brilliance without any warning.

You cannot help glancing at Cih every time you look up at Cassiopeia. Throughout the entire 20th century, astronomers watched the unpredictable variations in brightness that occur whenever the supergiant violently blows off the gaseous envelope around it. In 1937 it experienced a sudden inflow of energy and its brilliance doubled for a few weeks, reaching the luminosity of 8,000 suns. Although it has lost some of its brilliance since then, it still has sudden mood swings from time to time, and amateur astronomers keep it under constant watch.

2

ACHIRD
DOUBLE STAR

DISTANCE: **19.4 LY**
DIAMETER: **1.5 MILLION KM**
LUMINOSITY: **1.2 SUNS**

Not far from Cih shines Achird, a star strongly resembling the Sun, but with one difference: it is a double star.

Achird is relatively easy to see with the naked eye, given that it is so close to Earth — only 19.4 light-years away. It has a small companion that cannot be seen even through binoculars. But the couple are visible through an amateur telescope with a magnifying power of 100. Achird appears quite yellow, and its companion is reddish. This smaller red star is a dwarf, 30 times less brilliant than Achird itself, and it orbits around its partner over a period of 500 years. The two stars are separated by some 10 billion kilometers.

3

CASSIOPEIA CLUSTER
STAR CLUSTER M52

DISTANCE: **5,000 LY**
DIMENSIONS: **15 LY**

M52 is the brightest and most interesting of the numerous star clusters in the constellation Cassiopeia.

The Cassiopeia Cluster is beyond the reach of the naked eye but easy to view through binoculars, and looks like a small cloud right in the middle of the Milky Way. This sight becomes truly interesting through a small recreational telescope, when it appears as a bevy of hundreds of beautiful stars. Most of these stars are blue giants, all less than 100 million years old. At the center of M52 the concentration of stars is quite impressive: the giant stars there are separated from each other by less than one light-year.

PERSEUS

THE LATE NIGHT SKY

Perseus is quite easy to recognize in the northern sky, between the bright Capella in Auriga and the well-known W in Cassiopeia. The constellation draws two immense branches in the sky, which converge on a lovely group of stars dominated by Mirfak, the brightest star in the region. Also, by watching the constellation Perseus every night, you will have a front-row seat to observe Algol, "the demon," a star with a regularly changing sparkle.

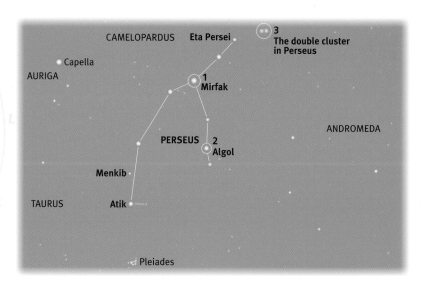

CAMELOPARDUS
Eta Persei
3 The double cluster in Perseus
Capella
AURIGA
1 Mirfak
ANDROMEDA
PERSEUS
2 Algol
Menkib
TAURUS
Atik
Pleiades

1

2

3

MIRFAK
SUPERGIANT STAR

DISTANCE: 500 LY
DIAMETER: 70 MILLION KM
LUMINOSITY: 5,000 SUNS

Mirfak sits in the very heart of Perseus, dominating a splendid group of about 30 stars that belong to the same cluster and are all about the same distance apart.

Mirfak is visible both to the naked eye and through binoculars. It is a supergiant that will soon come to the end of its brief existence in the dazzling flash of a supernova. Despite its remoteness, this monster is so gigantic that its diameter could only be measured using huge astronomical instruments such as interferometers. This blazing sphere of gas is 50 times bigger than our own star, the Sun. If it were to replace the Sun in the center of our Solar System, its fires would flame out as far as the orbit of Mercury.

ALGOL
VARIABLE STAR

DISTANCE: 93 LY
DIAMETER: 3.5 MILLION KM
LUMINOSITY: 100 SUNS

Algol is famous for its spectacular twinkling. Every two days and 21 hours, its brightness changes by a factor of three.

Its periodic variations in brightness are visible even without optical aids, making Algol the best example of what astronomers call eclipsing binaries. It is comprised of a giant star 100 times brighter than the Sun, and another giant four times brighter than the Sun, which revolves around its partner every two days and 21 hours. When observed from Earth, the orbital plane of the two stars can be seen almost in profile. In fact, every time the companion passes in front of the brightest star, it eclipses it for about 10 hours.

DOUBLE CLUSTER IN PERSEUS
STAR CLUSTERS NGC 869 AND 884

DISTANCE: 7,600 LY
DIMENSIONS: 70 LY

Toward the northwest corner of the constellation lies the double cluster in Perseus, probably one of the richest regions in the entire Galaxy.

These two clusters less than 10 million years old lie in the middle of the Milky Way. They are easy to observe under a clear country sky with the unaided eye and through binoculars. However, it is a real treat to observe them through a 300- to 400-mm telescope. Observers throughout history have been enthralled by the spectacle of these two groups, each containing several hundred stars and dominated by the sparkle of dozens of supergiants. Despite their remoteness, it is easy to make out the blue or red shades of these supergiants, some of which shine as brightly as a thousand suns.

TUCANA

THE LATE NIGHT SKY

The instruments of astronomers in the southern hemisphere are not pointed toward Tucana almost every night only to observe its faint, separate stars. They are more interested in the fact that this small easy-to-spot southern constellation, not far from the bright star Achernar in Eridanus, houses one of the most beautiful star clusters and, above all, the Small Magellanic Cloud, a satellite galaxy of the Milky Way. Despite its distance, this galaxy offers a wonderful sight both to the naked eye and through binoculars.

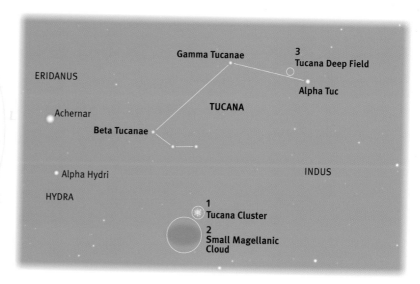

① TUCANA CLUSTER
STAR CLUSTER NGC 104

DISTANCE: 16,000 LY
DIMENSIONS: 200 LY

Not far from the Small Magellanic Cloud is the Tucana Cluster, prized, along with the famous Omega Centauri, for being one of the most beautiful star clusters in the entire sky.

To the naked eye, the Tucana Cluster looks like a small hazy star; however, when observed through a recreational telescope, it offers a sight of rare beauty. Through a small telescope, a few dozen very fine stars can be seen, crowded closely together. A large telescope reveals thousands of stars. Overall, the Tucana Cluster probably contains over a million stars, all created at the same time, about 10 billion years ago.

② SMALL MAGELLANIC CLOUD
IRREGULAR GALAXY

DISTANCE: 220,000 LY
DIMENSIONS: 20,000 LY

The light we receive today left this galaxy about 220,000 years ago. Yet, on the scale of the universe, this small satellite galaxy of the Milky Way is quite close to us.

Second only to the Andromeda Galaxy, the Small Magellanic Cloud is the most distant celestial object at the threshold of visibility to the naked eye. Seen through binoculars, it is beautifully related to the nearby Tucana Cluster. When observed through a very powerful telescope, the Small Magellanic Cloud reveals over a million stars as bright as Sirius. The total brightness of this galaxy equals close to a billion suns, and its mass is greater than 10 billion suns combined.

③ TUCANA DEEP FIELD
HUBBLE DEEP FIELD SOUTH

DISTANCE: 12 BILLION LY

This is a region empty of stars, in the vicinity of Alpha Tuc. The Hubble Space Telescope gazed into its unimaginable depths to explore the early universe.

Nothing can be seen in this direction of the sky, even with a large telescope. The Hubble Space Telescope was directed for an entire week into the Tucana Deep Field, a region of the galaxy transparent enough to view the most distant celestial objects, in order to obtain an image of the very remote universe as it was a few hundred million years after the Big Bang. Hubble counted 3,000 galaxies in this cosmic "keyhole," some of which are located over 12 billion light-years away.

DISCOVERING THE SKY

Modern civilization has forced the stars to withdraw from our skies. In recent decades, population centers have become increasingly large and dense, and the skies around them are more and more polluted with extraneous night light. While the moon, the planets and a small selection of bright stars can still be seen in the city, smaller stars, distant clusters, nebulas and galaxies have been eclipsed by excessively bright urban lighting.

Choose a clear sky

The Concise Atlas of the Stars was designed to be used under preserved skies. While the brightest constellations can still be seen from downtown building balconies, most of the celestial objects described in this book will be much better appreciated under a clear sky in open country or in the mountains. However, even in places far removed from city pollution, the most distant objects may not come out of hiding on the first try. First of all, the night you choose for observing must, of course, be a clear one without any trace of haze or clouds. Then, most importantly, the stargazer must patiently become acclimated to the night. The pupil of the eye opens fully—1/4 inch (5 and 7 mm) in diameter, depending on age—only after about 20 minutes of being in the dark. In other words, before trying to observe a dim, distant object, you should wait until your eyes have become well accustomed to the darkness. Also try to avoid extraneous ambient lighting, such as the lights of a nearby house or even the bright glare of a lighter or flashlight.

Choose the best conditions

To consult this atlas during observations, just attach a red filter to a flashlight. Red light—through red plastic film or nail polish, for example—does not affect the sensitivity of the eye to darkness, provided that the light is dim. Observing the sky is more than just pondering what lies in the heavens. It takes real concentration to fully perceive the subtle differences in the colors of the stars or the delicate nuances of the nebulous veils. So, to enjoy stargazing you will need to be absolutely comfortable. When viewing with the naked eye, lie right down on a blanket or sleeping bag when the constellation you are searching for is close to its zenith. If you are looking at the sky through binoculars, a deckchair is probably your best bet.

What equipment to use?

Finding constellations and observing the Milky Way with all of its rich discoveries can only be done with the naked eye. No optical instrument offers a field of vision wide enough to take in the Big Dipper, Andromeda or Sagittarius in one single shot. A 10x50 pair of binoculars (50 mm, about 22 inches, in diameter with a magnifying power of 10) offers the best balance between magnification and luminosity. By bringing into view celestial objects that are 100 times fainter than the faintest star visible to the naked eye, binoculars enable you to perceive, as small milky clouds, a very large number of star clusters, nebulas and even a few galaxies. They also provide a better view of the color of the stars than the naked eye. Observing the most immense and the most brilliant celestial bodies, such as the Omega Centauri or the Andromeda Galaxy, through binoculars is a real delight.

Finally, telescopes with a magnifying power between 50 and 300 allow you to really zoom in on the stars. However, the disadvantages are a tiny field of vision and monocular observation, which is less natural than the binocular vision available with binoculars. Double stars can only be seen with a telescope, and the richness and complexity of star clusters, nebulas and galaxies can only truly be appreciated when seen through instruments with a diameter of 100 to 500 mm (about 4 to 20 inches). Under a clear mountain sky, a large telescope capable of revealing the colors of the delicate veils of the nebulas and showing the spiral arms of the galaxies offers a literally infinite field of study.

Most of the celestial objects described in this atlas are visible to the naked eye, through binoculars or through a small amateur telescope.

PHOTOGRAPHING THE CONSTELLATIONS

Over the past few years, Akira Fujii (see box on next page) has acquired many admirers. Numerous recreational astronomers worldwide follow the Japanese photographer on his celestial treks to emulate his work.

With a simple camera

Nowadays, anyone can design their own collection of captured constellations, thanks to progress in optics, photographic emulsions and image processing software. While you may have no intention of rivaling the best astrophotographers, it is possible to learn how to photograph constellations without having to purchase any astronomical equipment. All you need is a simple 24x36 reflex camera and a camera stand. Most of the famous constellations fit into a regular 35- to 50-mm lens field. To increase your chances of success, set your equipment up on a clear night in an observation site free of any light. Then choose a bright constellation high in the sky in order to minimize problems with atmospheric absorption due to pollution or humidity in the air. Cygnus, the Big Dipper, Cassiopeia and Orion, or the Southern Cross in the southern hemisphere, are good places to start. The camera can be loaded with 800 to 1600 ISO color film, and the lens should be fully opened; that is, generally, between F:1.8 and F:2.8. All that remains is to set the shutter speed on the time exposure at B or T and to record several pictures at successive poses at 15 seconds, 30 seconds and 1 minute.

If everything works out well, the image will show all the bright stars in your chosen constellation. However, on these photographs the stars will not appear perfectly in place, but will have shifted a bit.

This is by no means the photographer's fault. This slight defect is caused by the movement of the Earth during photographic exposure.

The constellation Cygnus has been photographed here with a time exposure of 30 minutes on 100 ISO film, using a regular 58-mm lens stopped down to 1.2. The camera was mounted parallel to a small equatorial telescope.

With more sophisticated equipment

If you are looking to go further, to obtain images of constellations like those shown in this atlas, then you will need real astronomical equipment. In order to compensate for the rotating movement of the Earth, astronomers install their observation or photography equipment on special stands called equatorials. The main axis of an equatorial mount is power-operated and revolves around itself in 24 hours, like the Earth. For an equatorial mount to function accurately, its axis must be directed toward the celestial North Pole. Once pointed and turned on, the mount will naturally compensate for the apparent movement of the stars in the sky. Then anything is possible for the aspiring astrophotographer. To take the best image of a constellation, just test different types of film, lens apertures and even filters. The simplest method is to install your camera parallel to the tube of the refractor or telescope used to observe the sky. The astronomical instrument then serves as a simple viewer to be sure that the equatorial mount guide is perfectly set.

Watch for light pollution

It is perhaps worth repeating that photographing the constellations does not require any special optical instruments. A regular lens is more than enough for most snapshots. You will obtain excellent results from 100 to 400 ISO color films with lens apertures of F:1.8 for time exposures of 15 to 30 minutes. However, if the site where the pictures are taken lacks quality, you may be in for an unpleasant surprise when you have your film developed. Light pollution too often causes the sky in the background of an image to appear too bright, often milky or slightly green. There are two current methods for compensating for these drawbacks. The first technique is to confront the problem right when the shot is being taken. Expose part of the image — let's say, for example, from 10 to 15 minutes out of the total 30 minutes of exposure — through an interference filter placed in front of the lens. There are numerous types of interference filters on the market that let the light of the stars and nebulas pass through while blocking out artificial light from urban lighting. The second method is to scan and then process the image of the constellation after the shot is taken: image-processing software enables you to correct the color balance in photographs and to eliminate most of these bothersome sky backgrounds. Finally, the most efficient method is to use a combination of these two techniques: filtering and image processing.

AKIRA FUJII, PHOTOGRAPHER OF THE UNIVERSE

All of the full-page photographs of constellations presented in this atlas were taken by Akira Fujii. The celebrated Japanese amateur astronomer is the first to have dedicated himself to the remarkable and demanding task of completely photographing the sky, constellation by constellation.

Science and magic

Of course, the entire sky has been recorded by professional observatories over the past five decades. Astronomers already have a complete map of the sky with all of its wavelengths, from visible light to infrared light to radio-frequency radiation. But the sky proposed by Akira Fujii is on a human scale. It is a sky we can simply admire, with our noses in the air, on a perfectly clear night. That's what makes these photographs so magical and at the same time so familiar. Akira Fujii is indeed the first astronomer to have adapted classical photography techniques to astronomical imagery. The field of vision of his cameras, and the great number of stars visible in his photographs, portrays a night sky that is accessible to anyone, with the unaided eye or a set of binoculars. During the day, Akira Fujii is an illustrator, photographer and writer. In Japan, children love his works, which blend popular science and fiction. At night, he observes and photographs the universe with his telescopes. In order to be able to document the two celestial hemispheres, the astrophotographer has built two observatories, one in the Japanese Alps not far from Mount Nasu, and the other in the Australian desert.

Akira Fujii and his 300-mm telescope, installed in a dome of his Mount Nasu observatory. Akira Fujii has photographed all of the constellations in the northern and southern skies.

GLOSSARY

Big Bang: An event comparable to a gigantic dilatation of the universe over time that apparently led to its creation. Specialists today break down this dilatation, also called "expansion," into four main steps: the first, the "particle era," lasted only one second; the second phase, the "nuclear era," corresponds to about 100 seconds; the third, the "radiative-material era," represents 300,000 years! Finally, the fourth phase is called the "stellar" phase. About 15 billion years have elapsed from the creation of the universe to the present date. See also **Expansion.**

Binary or Double Star: A system of two stars connected by their mutual gravitational field. Each star may itself be a double star. The main star is brighter than the second, which is called a "companion."

Black Hole: A region of the universe where the gravitational field is so intense that nothing, not even light, can get out. Nothing inside it can communicate with the outside by any means whatsoever.

Blue Giant Star: A young star, with three to five times the Sun's mass and 100 times its luminosity. See also **Sun.**

Blue Supergiant Star: A very rare type of star in the galaxy. It is young, massive, hot (its surface temperature is about 30,000°C) and bright (it shines more brightly than 10,000 suns). See also **Sun.**

Celestial/Heavenly Body: Any natural object in the sky (planet, comet, star, galaxy, for example).

Cluster: A group of stars from the same galaxy (star cluster) connected by their mutual gravitational force. See also **Galaxy Cluster, Globular Cluster.**

Constellation: A group of stars, appearing to be in close proximity, that draw an arbitrary shape in the sky. Ancient civilizations have generally named constellations after an animal or hero of ancient mythology. In 1922, the International Astronomical Union set up boundaries in the sky and divided it up into 88 constellations. See pages 6 and 7.

Eclipsing Binary Star: A system of two stars that pass alternatively one behind the other, thereby hiding each other to an extent that reflects the orbit angle and the diameter of the two stars.

Electromagnetic Radiation: A form of energy characterized by the existence of closely associated variable electrical and magnetic fields, that spreads through empty space at a constant speed of about 300,000 kilometers per second. Light is an example of electromagnetic radiation.

Exoplanet: A planet revolving around a star other than the Sun. Also called an extra-solar planet.

Expansion: The dilatation of the universe over time. See also **Big Bang.**

Galaxy: A gigantic group of stars and interstellar material (gas and dust) held together by gravitation (mutual attraction). The universe has tens of billions of galaxies.

Galaxy Cluster: A large grouping of ten to several thousand galaxies. See also **Galaxy.**

Globular Cluster: A very dense, compact cluster containing hundreds of thousands of very old stars (about 15 billion years old), remnants of our galaxy's formation.

Light-year: A unit of distance (LY) equal to the distance that light or any other electromagnetic radiation travels in one year. Light travels at 300,000 kilometers per second, therefore one light-year equals 10,000,000,000,000 kilometers.

Milky Way: This is our galaxy. It is shaped like a disk measuring 100,000 light-years in diameter and is 10,000 light-years thick, as seen in profile.

Multiple Star: A system of at least three stars connected by their mutual gravitation.

Nebula: A vast cloud of gas and dust around a star or between stars.

Neutron Star: An extremely compact, small star representing a massive star's last stage of evolution, composed essentially of neutrons, the elementary particles that make up the nucleus of an atom.

Orbit: A curve that describes the path of a celestial body or a space vehicle around an object under the influence of its gravitational forces.

Planet: A celestial object not luminous itself that revolves around a star.

Primordial Galaxy: About 50 to 100 million years after the Big Bang, the very first stars condensed into the very first galaxies. Astronomers are beginning to observe these very remote primordial galaxies.

Red Dwarf Star: A small, faint star with little mass, whose life expectancy is several tens of billions of years.

Red Giant Star: A star with little mass, a hundred times brighter than the Sun, but at the end of its life. Its gas envelope has dilated and cooled. After expelling their envelopes, red giant stars become white dwarfs. See also **White Dwarf Star, Sun.**

Red Supergiant Star: A massive star reaching the end of its existence. It is gigantic, cold and not very dense, and ends its life cycle by violently exploding as a supernova. See also **Supernova.**

Solar System: A group formed by the Sun and the bodies that gravitate around it, including nine planets (Mercury, Venus, Earth, Mars, Jupiter, Saturn, Uranus, Neptune and Pluto, some of which have satellites and rings), as well as a multitude of asteroids and comets. It is located at 30,000 light-years from the center of the Milky Way Galaxy, around which it revolves over about 200 million years.

Spiral Galaxy: A galaxy whose bright stars and nebulas are spread out around a central core in spiral-shaped arms. Most galaxies have two symmetrical arms, but some of them have four, which is the case with our galaxy, the Milky Way.

Star: A very hot, gaseous body that produces and emits light.

Sun: A star with a fairly modest size and brightness, born about 4.5 billion years ago, located only 150 million kilometers from Earth. Its surface temperature is about 6,000°C. Its mass is about 330,000 times greater than the Earth's.

Supernova: The cataclysmic explosion of a massive star that has reached the end of its life. Supernovas are spectacular but very rare occurrences; only three occur per century in galaxies similar to ours. See also **Red Supergiant Star.**

Variable Star: A star whose variations in one or several features (luminosity, volume, temperature) are visible over time.

White Dwarf Star: A small, very compact star representing the last stage of evolution of stars with little mass. Its surface temperature is relatively high, and its luminosity is very faint (about one thousandth of the brightness of the Sun). See also **Red Giant Star.**

All 88 constellations are listed alphabetically and mapped on pages 6–7.

INDEX

Photo Credits

p.8-9: Nasa/STSCI/Ciel & Espace — p.10-11: David Malin/AAO/Ciel & Espace — p.12-13: S. Numazawa/APB/Ciel & Espace — p.14-15: David Malin/AAO/Ciel & Espace — p.16: g,m,d, Akira Fujii/Ciel & Espace — p.17: Akira Fujii/Ciel & Espace — p.18: g, Akira Fujii/Ciel & Espace • m, Jerry Lodriguss/Ciel & Espace • d, Nasa/STSCI/Ciel & Espace — p.19: Akira Fujii/Ciel & Espace — p.20: g, Akira Fujii/Ciel & Espace • m, Bill et Sally Fletcher/Ciel & Espace • d, Akira Fujii/Ciel & Espace — p.21: Akira Fujii/Ciel & Espace — p.22: g, Akira Fujii/Ciel & Espace • m, P. Riepe/B. Schroter/Ciel & Espace • d, John Gleason/Ciel & Espace — p.23: Akira Fujii/Ciel & Espace — p.24: g,m, Akira Fujii/Ciel & Espace • d, ESO — p.25: Akira Fujii/Ciel & Espace — p.26: g, Akira Fujii/Ciel & Espace • m, d, David Malin/AAO/Ciel & Espace — p.27: Akira Fujii/Ciel & Espace — p.28-29: David Malin/AAO/Ciel & Espace — p.30: g, Akira Fujii/Ciel & Espace • m, David Malin/AAO/Ciel & Espace • d, Akira Fujii/Ciel & Espace — p.31: Akira Fujii/Ciel & Espace — p.32: g,m, Akira Fujii/Ciel & Espace • d, Nasa/STSCI/Ciel & Espace — p.33: Akira Fujii/Ciel & Espace — p.34: g, Akira Fujii/Ciel & Espace • m, Bill et Sally Fletcher/Ciel & Espace • d, Akira Fujii/Ciel & Espace — p.35: Akira Fujii/Ciel & Espace — p.36: g,m, Akira Fujii/Ciel & Espace • d, Bill et Sally Fletcher/Ciel & Espace — p.37: Serge Brunier/Ciel & Espace — p.38: g,m, Akira Fujii/Ciel & Espace • d, David Malin/AAO/Ciel & Espace — p.39: Akira Fujii/Ciel & Espace — p.40: g,m,d, Akira Fujii/Ciel & Espace — p.41: Akira Fujii/Ciel & Espace — p.42: g, Akira Fujii/Ciel & Espace • m, Aura/NOAO/Ciel & Espace • d, Aura/NOAO/Ciel & Espace — p.43: David Malin/AAO/Ciel & Espace — p.44: g, Akira Fujii/Ciel & Espace • m, N. Solee/Ciel & Espace • d, ESO/Ciel & Espace — p.45: Akira Fujii/Ciel & Espace — p.46: g, m, Nasa/STSCI/Ciel & Espace • d, Ciel & Espace — p.47: David Malin/AAO/Ciel & Espace — p.48: g,m, Akira Fujii/Ciel & Espace • d, Nasa/STSCI/Ciel & Espace — p.49: Akira Fujii/Ciel & Espace — p.50: g,m, Akira Fujii/Ciel & Espace • d, Bill et Sally Fletcher/Ciel & Espace — p.51: Akira Fujii/Ciel & Espace — p.52-53: John Gleason/Ciel & Espace — p.54: g,m, Akira Fujii/Ciel & Espace • d, N.A. Sharp/AURA/NOAO/Ciel & Espace — p.55: Akira Fujii/Ciel & Espace — p.56: g,m, d, Akira Fujii/Ciel & Espace — p.57: Akira Fujii/Ciel & Espace — p.58: g,m, ESO/Ciel & Espace • d, Nasa/STSCI/Ciel & Espace — p.59: Akira Fujii/Ciel & Espace — p.60: Alain Cirou/Ciel & Espace — p.61: g, Serge Brunier/Ciel & Espace, d, Alain Cirou/Ciel & Espace